THE SCIENCE OF WEATHER

BIG IDEAS: LOW INTERMEDIATE

ALEX SEMAKIN

WAYZGOOSE PRESS

CONTENTS

Edited by Robyn Brinks Lockwood

Cover design by GetCovers.com

SUPPLEMENTARY MATERIALS

For Big Ideas downloadable learning tools for students and teachers, go to
https://www.wayzgoosepress.com/downloads/.

INTRODUCTION

When you read an article or story, you have a conversation. The writer shares information, experiences, and ideas—but you, the reader, have your own ideas. When you read, you compare your experience and knowledge with the writer's ideas. Then you make decisions. Do you agree with the writer? Can you use the information to be healthier or more successful in your job, for example? Do you feel like the writer understands your life? Or do you learn about someone with a different point of view?

Because reading is a conversation, every reader experiences a text differently. When you read something interesting, you often want to talk about it. You want to share a similar experience, or you may want to argue. Maybe your friend understands the text in a different way. When you listen to your friend, you have a third set of ideas and experiences to compare to your own world view.

Big Ideas is designed to start interesting conversations between readers and writers, but also between readers and other readers. In this book about weather, you'll learn about weather and meteorology, the science of the atmosphere and forecasting. The chapters explain why we need to study the weather and introduce some people and tools in the history of meteorology. You will also learn about some weather records, climate change, and forecasting. Meteorology is important for forecasting or predicting weather, so we know what to wear or prepare before we start our day. It is also important to study trends in weather to understand the impact on climate change and the future of the world. There are always new and interesting questions and topics to explore.

While you learn about the science of weather, *Big Ideas* is also helping you develop language skills. Because our focus is on providing a positive reading experience, more than 90 percent of the words in this book are among the most common 2000 words in the English language. These are called "high frequency words." High frequency words appear over and over again in speaking and writing.

You might think it will be easy to learn high frequency words, and it is true that many words are easy. Content words, such as *tree, house, eat, drink*, and *blue* put a picture in your mind. They represent things you can see and name. They often have one meaning, and you can translate them easily.

However, many high frequency words change their meaning when they partner with other words in collocations. *Stay* is an example. When we say, *I stayed home yesterday*, then *stay* has a different meaning from *Let's try to stay awake all night* or *Stay away from the cookies. I'm saving them for the party.* This flexibility shows that *stay* does not just have one meaning. It adapts to the words around it.

Fortunately, there is a method to learn the different meanings of collocations: read a lot. When you read, you see the words in different combinations, and you learn the meanings. This can happen naturally, but it will happen faster if you pay attention to words in groups. When you notice and highlight or copy word combinations, you can learn the different meanings.

You can also learn the grammar that goes with a vocabulary word. For example, you might see *educate* as a verb in *educate children*, *education* as a noun in *college education*, and *educational* as an adjective in an *educational experience*. Or, in another example, you can notice that some verbs are usually followed by a preposition, such as talk about or talk to, while others are followed by a noun as in *hear a bird*. These grammatical details are hard to hear in spoken English, but they are easy to spot in a written text.

While vocabulary has a strong relationship with grammar, grammar has a strong relationship with sentences. In order to give you a positive reading experience, we have worked hard to provide easy-to-read sentences. We use grammar from intermediate levels, and we reduce synonyms and idioms. Our goal is to keep the big ideas about weather but present them in simple language. Since there are a lot of science and weather words in this book, we have a vocabulary section at the beginning. You can study the words before you start reading, or come back to that section later if necessary. Or, of course, you can check unknown words in your dictionary.

VOCABULARY FOR WEATHER

How you handle vocabulary when you read is your choice. However, we suggest reading through this list of words once or twice before you begin. These are important words for understanding and talking about weather, and you will see them many times in this book. If you need to learn them, you can come back to this section and read the definitions again, or make flashcards with the vocabulary and teach them to yourself. You can also check them in a bilingual dictionary.

atmosphere: All the air that surrounds the earth. Thanks to the atmosphere people can breathe and there is life on the planet.

data: Information; usually in the form of numbers. For example, weather stations collect all types of weather data, which meteorologists use to make forecasts.

density: How "thick" the air is. Dense air has a large amount of gases in a small space. The air at sea is denser than the air high in the mountains.

drought: A long period of dry weather. During a drought, rivers and lakes may dry up and people may not have enough water to drink.

evaporate: To turn into steam. Any water that is not in a closed container begins to evaporate. It evaporates faster when it is hot than when it is cold.

eye (of a storm): The circular area about 40-50 kilometres wide at the very centre of a tropical cyclone. The weather is usually calm in the eye of a storm.

forecast: Information about what the weather will be in the next hours, days, weeks, or months. Meteorologists study weather and make weather forecasts. Forecast is also used as a verb: *I can't forecast the weather.*

gale: A very strong wind. Gales often blow from the sea. They are a frequent weather event in the United Kingdom.

hail: Balls of ice that fall from the sky like rain. A heavy thunderstorm can turn into a hailstorm. Hail can destroy crops and damage objects.

heat wave: A period of unusually hot weather that lasts more than two days.

humid: Warm and damp (with water in the air). In many seaside areas, the weather is often humid. It is harder for some people to breathe and exercise in humid weather. **Humidity** is the amount of water in the air. For example, if the humidity today is 30 percent, it's a dry day; if the humidity is 90 percent, it's a wet day; 100 percent humidity means it is foggy.

hurricane: A powerful storm with very strong winds. It comes from the ocean and can bring a lot of destruction to coastal areas.

measure: To find the size, quantity, or level. For example, people measure temperature to know how warm or cold it is. Temperature is usually measured in Celsius (C) or Fahrenheit (F).

observe: To watch something carefully for some time, especially to know more about it. For example, biologists observe animals and plants; astronomers observe planets and stars; meteorologists observe weather. The noun form is **observation.**

overcast: completely or almost completely covered with clouds. In meteorology, an overcast sky or an overcast day means that clouds cover more than 95% of the sky.

precipitation: Anything that falls to the earth from clouds, such as rain, snow, sleet, or hail. Meteorologists measure the amount of precipitation in inches or millimetres/centimetres.

predict: To say what will happen in the future. When talking about weather, *predict* means the same as *forecast*. The noun form is *prediction*. A weather prediction is the same as a weather forecast.

pressure (air pressure, atmospheric pressure): The force with which all the air above weighs down on people and objects. Pressure changes depending on the weather and on the height above the sea level.

record:

1. (noun, pronounced *REcord*) A written note about a fact or event. Records are kept to look at and use in the future. People have kept weather records for centuries.
2. (verb, pronounced *reCORD*) To keep notes about facts or events. People have recorded weather for centuries.
3. (noun, pronounced *REcord*) The highest or lowest level of something that has ever been reached. Weather records are information about the most extreme circumstances, such as temperature, wind, or precipitation in history of observation.

scale: A range of levels or numbers used for measuring something. For example, the Celsius scale is used in most countries to measure temperature.

sleet: A kind of precipitation that forms when the temperature is near freezing, and the falling snow melts and then freezes again before reaching the ground. You can tell sleet from snow because sleet bounces when it hits the pavement. You can tell sleet from hail because hail is larger and happens when the weather is much warmer.

smog: The kind of fog that forms over cities where humidity is high and the air is very polluted. It contains smoke and gases which make it difficult to breathe and are dangerous for people's health.

surface: The top of an area of water or land. For example, we walk on the surface of the earth; ice floats on the surface of rivers in spring.

temperature: How warm or cold something is, measured in degrees. When meteorologists talk about temperature, they usually mean air temperature–how warm or hot the air is at a certain place. Water temperature can be measured too.

tornado: A local storm with very strong winds that move in a circle. It looks like a long cloud with its bottom tip reaching the Earth. A tornado can pick up heavy objects and move them over long distances.

tropical cyclone: The same thing as a hurricane. The word *hurricane* is usually used for cyclones that hit North America. Other places refer to hurricanes as *tropical cyclones*. In the Northwest Pacific Ocean, they are also called *typhoons*.

PART 1

WHY DO WE NEED METEOROLOGY?

Clouds seen from an airplane

Meteorology is the branch of science that studies the atmosphere. One area meteorology focuses on is *forecasting,* or predicting, the weather. In this short opening section, you will find out why meteorology is important for everyone.

~

On the 2nd of February every year, crowds of people come to the small town of Punxsutawney, Pennsylvania, in the US to watch a groundhog named Phil come out of the hole he lives in underground. Why do they do that? They believe that Phil can predict the future. If the day is sunny and Phil sees his shadow, it frightens him and he runs back into his hole. People think that it means winter will last six more weeks. If the day is cloudy and Phil doesn't see his shadow, he stays out and looks for food. This means spring will begin soon.

This 2nd February event became famous around the world thanks to the 1993 film *Groundhog Day*. But are Phil's forecasts reliable? Not really. In fact, only 39% of Phil's past predictions have come true. However, that doesn't stop people from coming to watch the groundhog every year and believing–or pretending to believe–in his predictions.

A groundhog

For thousands of years, humans have wanted to know what weather to expect in the next hours, days, weeks, and months. Knowing the weather helps us make plans for travel or outdoor events. It helps us make decisions about what to wear, whether or not to take an umbrella with us, or whether to stay away from trees and go inside. Farmers need to know the weather to successfully grow food. Pilots need to know the weather to fly safely. Film directors need to know the weather to plan filming outdoor scenes.

The science called *meteorology* helps. Meteorology is the

study of the earth's atmosphere. It tells us what happens between the surface of our planet and the highest clouds. Meteorologists observe and record weather and make weather forecasts.

Unfortunately, no forecast can be 100% reliable. Sometimes the radio predicts a sunny day, so you leave your umbrella or raincoat at home. Then it rains, and you get wet. Sometimes the opposite happens: your weather app predicts rain, so you carry your umbrella with you all day and never use it because the day is sunny and dry. Or your plane is delayed by fog without any fog in the forecast. Or the TV announces temperatures above freezing, but then rain freezes on the street, and you fall down and hurt your knee.

These are all reasons people complain about meteorologists and poor predictions. Some even joke that meteorology is the only occupation where you can be wrong all the time and still get paid. Is that fair? Probably not. The truth is that weather forecasting today is the best it has ever been. The simple fact that a weather forecast is often correct is a miracle of science, mathematics, and technology, and meteorologists keep looking for new ways to make even better forecasts.

It has taken mankind thousands of years to reach this stage.

Reflection

1. Can you think of some other occupations or groups who benefit from weather forecasts? Why do they need to know the weather?
2. How important are weather forecasts for you in your everyday life? Do you often check the weather forecasts? Why or why not?

PART 2

THE BEGINNINGS OF
WEATHER SCIENCE

An early instrument to measure humidity

Throughout history, there have been many important discoveries that have shaped the science of meteorology. In this section, you will read about people's earliest interest in weather and about the first observations and discoveries they made. You will also learn about some inventors who created some of the instruments used in meteorology.

1

LOOKING UP TO THE SKY: BEFORE METEOROLOGY BECAME A SCIENCE

We are all alive because our planet has an *atmosphere*. The atmosphere is the air that we breathe. The air is not staying in the same place. It is moving all the time. As it moves, it carries the temperature and humidity with it. Weather is simply the result of our atmosphere moving heat from one place to another. This movement is very important for all of us.

Have you noticed that many people start conversations by talking about the weather? Weather is one of the most popular subjects of casual conversations because weather affects our everyday lives. Weather connects people who otherwise have nothing in common. Whatever your age, occupation, or beliefs, you will sweat on a hot day, freeze on a cold day, and get wet on a rainy day. Today, just like thousands of years ago, people look up to the sky and wonder what surprises it will bring.

It is possible that when humans first began to talk, they talked about weather. At that time, nobody understood the real reasons for weather changes. People believed that gods or mythical creatures ruled the sky.

Upanishads, the Indian text from about 3,000 years ago, is the first known text where weather was discussed. The author tried to explain how clouds appeared, why it rained, and how seasons changed. Many years later, around 650 B.C., the Babylonians tried to predict weather for the first time based on cloud patterns and the stars.

Ancient Greeks were interested in weather changes too. The word *meteorology* comes from Greek. It was invented by Aristotle, the Greek philosopher. Around 340 B.C., Aristotle wrote a book called *Meteorologica*, which means 'the science of things high up.' The book described everything people knew at that time about weather and climate.

One of Aristotle's biggest achievements was describing the water cycle. He wrote about the three main stages of the water cycle. First, the sun heats the water on the surface of the earth and turns it into steam. Then the steam rises and becomes clouds. Finally, the water returns to the earth as rain or snow. Aristotle also guessed that rain water might go into the ground, flow underground, and then return to rivers and seas. However, Aristotle also believed in things that sound very strange to us today. For example, he did not believe that wind was moving air. He believed that west winds were cold because they blew from the sunset. With such misunderstandings, it was not possible to accurately predict weather.

Different peoples around the world tried to predict weather before modern meteorological science appeared. They had no advanced instruments, but they made predictions from careful observations.

Among those peoples were Native Americans. For example, the Chugach people in Alaska used weathermen, who spent some time every day looking at the sky. They observed clouds and predicted weather from the cloud shapes. At night, they looked at the stars. If the stars *twinkled* (changed their brightness) slowly, the weathermen said the next day would be fine and calm. If the stars twinkled quickly, they said the next day would be windy.

Further south, in what is now Mexico, Mayan people paid close attention both to the sky and to nature around them. They observed the behaviour of plants, insects, and animals to decide when to plant crops. The Mayans tried to make long-term forecasts. Each January, they observed the weather many times a day for 12 days. They believed that those 12 days were connected with the 12 months of the year ahead. They thought that the weather on a certain day showed what the weather would be in a certain month. They also put 12 piles of salt on a table in the evening of 31st December. Each pile represented a month of the year. If a pile fell down by the next morning, it meant that month would be humid.

We can laugh at some of those predictions and call them unscientific. Their forecasts were often based on beliefs and not facts. But if Native Americans continued to believe in them for a long time, maybe they weren't always wrong.

Consider, for example, this story about the Incas, the natives of modern Peru. Every June, a group of stars called the Pleiades, or Seven Sisters, appears in the southern sky. At the end of June, a group of Incan priests used to carefully observe the Seven Sisters from Korikancha, the main temple of the Incas in Cusco. They did that to predict the weather to decide when to plant crops. They believed that if the Seven Sisters looked large and bright, the year would be full of rain, and crops would grow well. If they looked small and not very clear, the priests predicted a year with less rain that would arrive late, so farmers should plant their crops later. Such observations continued for centuries.

Is it possible to simply watch a group of stars and predict the weather for many months in the future? Scientists wanted to know more about these traditional beliefs. Benjamin

Orlove, John C. H. Chiang, and Mark A. Cane studied these beliefs at the beginning of the 21st century. While researching the Incas' forecasts, they discovered a connection between high clouds and El Niño. El Niño is the warming of the ocean that takes place every several years around the equator on the west coast of South America. The scientists found that high clouds appear more often in El Niño years. The high clouds are so thin that people can't always see them during the day. At night, however, the Seven Sisters look less bright if the high clouds are there. In El Niño years, there is less rain, and it arrives later in September, October, and November. When the researchers compared the Incas' forecasts with the modern forecasts for the same area and the same amount of time, they were amazed. The accuracy of the ancient forecasts was 65%. The accuracy of modern forecasts was 55 to 60%. The Inca priests predicted El Niño better than meteorologists do today! Or at least, better than they did in 2002, when Orlove, Chiang, and Cane published their findings in *American Scientist*.

In China, people have been recording weather for about 3,000 years. Scientists have discovered thousands of historical diaries with daily weather data. The same person often recorded the weather every day for decades. The diaries recorded when it was hot or cold, when it rained or snowed, and when there was a thunderstorm or a dust storm.

Chinese history books explained the meanings of different types of clouds. Their authors tried to predict the weather from the clouds based on many years of observation. For example, *The Book of Songs*, the earliest collection of poetry, written between the 11th and 5th centuries B.C., noted, "If the clouds have the same colour and similar thickness, it will snow." It also noted that if the clouds looked like castles, there would be heavy rain and that if the clouds had black and red in them at the same time, there would be hail. Artists painted different kinds of clouds first on silk and later on paper.

The government of the Qing Dynasty, which ruled China from the 17th to the early 20th century, kept daily weather records. The data was put in two books. One book described general weather, and the other described precipitation. Another book, *Bai Yuan Xian San Guang Tu,* contained 132 pictures of clouds, with explanations next to each picture.

Meteorologists still use some observations from ancient Chinese books to make forecasts today.

Reflection

1. One reason ancient cultures studied the weather was to help their farmers. What are some other reasons weather forecasts were important for them?
2. Can you look at the sky and tell what the weather will be like later today or tomorrow? What are some observations or beliefs you know that people use to predict the weather?

2

TORRICELLI'S BAROMETER AND THE BEGINNING OF MODERN METEOROLOGY

Modern meteorology was born in the 17th and 18th centuries, thanks to some very important achievements by chemists and physicists of that time.

Robert Boyle and Jacques-Alexandre-César Charles discovered the laws of gas pressure, temperature, and density. Isaac Newton and Gottfried Wilhelm Leibniz developed calculus, an advanced type of maths. John Dalton discovered the law of partial pressures of gases. Joseph Black described the energy produced when water freezes or boils and when ice melts.

Because of all those advances in science, people could better understand what happened in the atmosphere. They began to solve the mysteries of weather.

The two most important inventions for the future of meteorology were the mercury barometer in the mid-17th century and the mercury thermometer in the early 18th century. Mercury is a liquid metal that is used to measure changes in temperature.

The inventor of the mercury barometer was the Italian scientist Evangelista Torricelli. Torricelli was born to a poor family. His parents didn't have money to pay for his education, but they noticed that he was very intelligent, so they sent him to his uncle Gian Francesco, a priest and a well-educated man. Gian Francesco realised that Evangelista was brilliant at mathematics. He began teaching the boy himself and then sent him to school. From 1626, Torricelli studied in Rome. He was taught mathematics and physics there by Professor Castelli, a former student of Galileo.

Torricelli admired Galileo and his work. One day, he decided to write a letter to Galileo describing his own ideas and interests. Galileo was very interested. He and Torricelli exchanged many letters. Just before Galileo's death in 1642, Torricelli visited him and spent several months living and working in Galileo's house.

Scientists before Torricelli thought that the air had no weight. Torricelli proved that we live in 'a sea of air'. All the air weighs down on us, like water in the sea weighs down on a diver. But air is less dense than water. This explains why a diver feels water pressure, but people on the surface of the Earth don't usually feel air pressure.

In 1643, Torricelli invented the barometer to measure atmospheric pressure. The word *barometer* comes from the Greek words for *measure weight*. The instrument used mercury. When the air pressure on the barometer was high, the mercury in it rose. When the pressure was low, the mercury fell.

After Torricelli's death, Blaise Pascal proved that air pressure becomes less when we climb up a mountain. The air pressure on top of a mountain is much lower than the air pressure by the sea. Can you guess where the average air pressure is the lowest on Earth? The answer is Mount Everest. Where do you think the average pressure is the highest? Near the Dead Sea, which is the lowest place on Earth. This information might make you think about airplanes. Airplanes fly higher than mountains. Why don't passengers feel sick when they fly, since the air pressure is very low? The reason is that normal pressure is created inside a plane so passengers will feel comfortable. This pressure is much higher than the air pressure outside the plane.

Scientists after Pascal realised that air pressure could be different at the same height, too. The difference depends on air density. When density is high, the pressure is higher and the air is colder. The sky is usually clear then, and the sun warms up the cold air in summer. When air density is low, the pressure is lower and the air is warmer. The sky is usually cloudy then. As Torricelli correctly guessed, winds always blow from the areas of high pressure towards the areas of low pressure. Using a barometer, meteorologists can predict changes that will affect the weather, so the invention of the barometer was very important for meteorology.

Reflection

1. Find your local weather forecast. What is the air pressure today? What is it expected to be tomorrow?
2. Do a little online research to find out what effect air pressure may have on people. Why is it important for people to know the pressure forecast?

3

THE THERMOMETER: FAHRENHEIT AND CELSIUS

The thermometer, which measures temperature, was another important instrument in the development of modern meteorology. The thermometer was invented by Gabriel Fahrenheit. Fahrenheit invented both the instrument and the temperature scale that is named after him.

Over the centuries, people have used many different scales to measure temperature. Some scales, such as Newton and Leiden, are no longer used. The Rankine scale is still used in engineering. The Réaumur scale is still used in food factories to measure the temperature of milk when cheese is made. The Kelvin scale is still widely in physics. But the best-known scales today are Fahrenheit and Celsius. Both measure temperature in degrees: degrees Fahrenheit (°F) and degrees Celsius (°C). These scales are used in weather reports and forecasts around the world.

Fahrenheit was an engineer and physicist. He was born in Poland to a German family. Like Torricelli, he was a talented boy, but unlike Torricelli, he had rich parents who sent him to a very good school. Unfortunately, they died before he could finish school. The people who took care of him after his parents' death paid for him to study at a business school in Amsterdam. But Fahrenheit didn't want to study there, and he didn't finish his degree. He left Holland and spent some years travelling around Europe. During his travels, he met with many scientists and instrument makers. He finally went to live in the Hague and worked as a glassblower, making different measuring instruments.

In 1708, Fahrenheit met Ole Rømer, the Danish scientist who discovered the speed of light. Rømer was making alcohol thermometers using his own temperature scale. Impressed, Fahrenheit invented a method to produce Rømer thermometers in large numbers. Fahrenheit continued to research and learned that with mercury, he could measure temperature more accurately than with other materials.

However, Fahrenheit needed a new scale for the new thermometer, so in 1724, he invented his own temperature scale. He took the lowest temperature he could achieve in his lab–the temperature of ice, salt, and water mixed–and made that temperature 0 (zero) on his scale. He then put a thermometer into the mouth of a healthy person, possibly himself, and called that temperature 96 degrees. No one knows why he chose that number! Then he divided the scale into single degrees from 0 to 96. On this scale, water froze at 32 degrees and boiled at 212.

In 1742, Swedish scientist Anders Celsius invented a different scale. Celsius called the freezing point of water 100 and the boiling point 0. A year later, French physicist Jean-Pierre Christin suggested turning Celsius's scale upside down. It is actually Christin's scale, with 0 for freezing and 100 for boiling, that we use today. However, since the original idea belonged to Celsius, the scale is named after him. The idea of connecting a scale to the states of water and using round numbers seems obvious now, but it wasn't at the time.

Celsius lived after Torricelli, so he knew about the effects of atmospheric pressure. He wondered if the temperature of freezing and boiling depended on pressure. He experimented and discovered that the freezing point did not change with a change in pressure. However, the boiling point did change. Water boiled at slightly lower temperatures at higher altitudes. Then he announced that the boiling point on his scale would be the temperature of water boiling in the average air pressure at the average sea level. That pressure is now known as *one atmosphere*.

For a long time in the 19th and 20th centuries, Fahrenheit was the main scale used around the world. Celsius's scale has become more popular than Fahrenheit's in most places because it is easier to understand and more convenient to use. Today, almost all countries measure temperature in degrees Celsius. The US is one of the few countries still using Fahrenheit. The UK changed from Fahrenheit to Celsius in 1962. Today, Celsius is used there in weather forecasts, but many British people still understand Fahrenheit temperatures. Fahrenheit is sometimes used in newspaper headlines to impress readers. To say that it is 100 degrees in the south of England sounds more shocking than to say it is 38 degrees, even though they are exactly the same.

Reflection

1. Do some online research to find out about one of the temperature scales that is no longer used. How did it measure temperature? Why do you think it is not used today?
2. Why do you think the Fahrenheit and Celsius scales are still popular?

PART 3

FROM OBSERVATIONS TO FORECASTS FOR EVERYONE

Measuring wind at an airport

Equipped with the new instruments, scientists began to organise their knowledge of weather. They described the types of clouds and created a wind scale. They started making forecasts to warn people about dangerous weather. In this section, you will find out how modern forecasting began and how information about weather finally reached everyone. You will get to know a few more famous people whose work shaped the science of meteorology.

4

LUKE HOWARD: THE GODFATHER OF CLOUDS

As you read in Chapter 3, the clouds were very important to the ancient Chinese in their weather observations and predictions. Today, clouds are still very important in forecasting. In fact, clouds and wind are the two main things that cause changes in weather.

The first person to give scientific descriptions of clouds was Luke Howard. Howard was a British chemist and businessman. He never became a professional meteorologist, but he fell in love with meteorology as a young child. More than anything, he loved the clouds. Howard kept detailed records of the weather in the London area every day from 1801 to 1841. Even though he was not professional, his writings changed the science of weather. Sometimes Howard is called the 'Father of Meteorology'.

In 1802 and 1803, Howard described and named several types of clouds. Therefore, Howard is also called the 'Namer of Clouds' and even the 'Godfather of Clouds' (a *godfather* is responsible for helping parents take care of a child). His names were based on what the clouds looked like and how they were formed. His 1803 book about clouds had many detailed drawings of clouds he drew himself. Meteorologists still use Howard's cloud names.

If it's daytime when you are reading this, look at the sky. Perhaps it is a grey day with an overcast sky, with the clouds covering the sky like one enormous blanket. These clouds are *stratus* clouds, from the Latin word for *sheet*. Stratus clouds are more common in winter than in the summer.

Some different types of clouds

If it's a warm, pleasant, and mostly sunny spring or summer day, you may see *cumulus* clouds, from the Latin word for *heap*, which is like a mountain or pile of clouds. These are often light and fluffy, like a pile of cotton. Cumulus clouds are signs of good weather. However, they may grow several kilometres both up and down, bringing rain and thunder.

The highest and lightest clouds are *cirrus* clouds from the

Latin for *hair* because they look like hair blowing in the wind. These clouds do not carry rain but may bring some light snow in winter. The sun can shine through cirrus clouds. It was very high cirrus clouds that changed the appearance of the stars and helped the Inca priests predict El Niño.

Howard's work inspired not only scientists but also poets and painters. The German writer and poet Goethe wrote several poems celebrating Howard's discoveries. The English poet Shelley wrote his poem *The Cloud* thinking about Howard's cloud types. The English painter Constable created many paintings of skies with clouds.

Howard lived and worked in Tottenham, London. In 2018, the football club Tottenham Hotspur named two viewing areas in their new stadium Stratus East and Stratus West, after Howard's clouds, to show respect for the great man and his discoveries. When people are not watching the game, they can see far across London from these viewing areas. Of course, they can see clouds in the sky as well.

Howard described the most common types of clouds. The clouds you see above you are probably one of Howard's types. But if you are very lucky, you can see some rare kinds of clouds, too. For example, *lenticular* clouds look like UFOs (unidentified flying objects). Seeing them makes some people think that visitors from another planet have arrived. They may post photos of those clouds to prove that they have seen alien spaceships. *Kelvin Helmholtz* clouds look like ocean waves. *Mammatus* clouds look like corn. Beautiful *nacreous* clouds have many colours and look like the Northern Lights.

Reflection

1. Howard was not a meteorologist, but he fell in love with meteorology. What are some reasons a child might fall in love with meteorology?
2. Look at the sky. What kinds of clouds are there today? What do they look like? If you can't see any clouds today, check some photos that you took outside, and identify the clouds you see there.

5

FRANCIS BEAUFORT AND HIS WIND SCALE

The name of Sir Francis Beaufort is forever connected with wind.

Beaufort was born in Navan, Ireland in 1774. His father, Daniel Augustus Beaufort, was a well-known geographer who created and published the most detailed map of Ireland. Beaufort grew up in Wales and Ireland. When he was 13, he left school and went to sea, but he never stopped his education. He taught himself so well that he later became friends with some of the greatest scientists and mathematicians of his time.

When Beaufort was 15, he was a young sailor in the British Navy. On one voyage, his ship sank, although he survived. The reason the ship sank was because the sea map the Navy was using was wrong. After that, Beaufort developed an interest in making sea maps, which became one of his lifetime achievements. Some of his maps are still used today, over 200 years after he made them! But his greatest achievement was the Beaufort wind scale.

At age 16, Beaufort became interested in weather. He started a diary where he wrote his weather observations every day. He kept the diary until the end of his life. On some days he wrote notes in the diary every two hours. Because he wrote so frequently, he needed a quick way to record different weather conditions. He invented a one-letter code for each type of sky, cloud condition, and precipitation. For example, *b* meant *blue sky*, *o* meant *overcast*, *r* meant *rain*, and *t* meant *thunder*. If he wrote *s.c*, it meant *snow and some clouds*. If he wrote *g.r.q*, it meant *gloomy* (dark), *rainy*, and *strong wind* (called a *squall*). Modern meteorology codes are still based on Beaufort's codes.

His greatest achievement, however, was the Beaufort wind scale. Beaufort felt that his descriptions were not clear enough. Something else was needed to make them clearer. In 1805 and 1806, during a voyage to South America, he created a wind force scale. From then on, he wrote a number from the scale next to each weather code in his diary. The scale, with some changes, is still used by meteorologists around the world today and is known as the Beaufort scale.

The initial Beaufort scale was not exact. It did not give the wind speed. The numbers were based on what the wind did to ships, the sea, and objects on land. In other words, they were based on the wind's effects, which allowed Beaufort to measure wind without using instruments. For example, Beaufort number 0 means total calm: the sea is like a mirror, and smoke rises directly up. Number 3 means there is a gentle breeze: there are some waves with white tops, and leaves on trees are moving all the time. Number 7 means there is a gale, or high wind: the sea waves are high, and it is hard for people to walk against the wind. Number 10 means there is a storm: the whole sea turns white, the waves are very high, and it is harder to see. The wind brings down some trees and blows the roofs off some buildings. The strongest wind, a 12 on the Beaufort scale, means a hurricane. The air and sea are all white, the waves are extremely high, and the wind destroys almost everything on land.

Beaufort's scale was used for the first time on the voyage of the *Beagle* from 1831 to 1836. This voyage is famous because one of the ship's passengers was Charles Darwin, who was sailing to South America to research animals and plants for his book. After that voyage, more people started using the Beaufort scale. In 1838, all ships in the British Navy were ordered to record the force of wind in Beaufort numbers. For each Beaufort number, ships had to use a specific number of sails. Much later, in 1926, each number on Beaufort's scale got a speed range. Those ranges have never changed. For example, number 3 means the wind is blowing 12 to 19 kilometres per hour. In 1946, numbers up to 17 were added to the scale for hurricane-force winds.

Francis Beaufort was not the first man who studied wind. Almost 100 years before him, George Hadley explained the direction of winds around the equator. He did this because it was important for European trade ships sailing to the new continent—North America. Hadley discovered that the hot air near the equator rises to a height between 10 and 15 kilometres. It then moves quickly towards the South and North Poles and begins to cool. The cooled air begins moving east and then down. As it pushes down, it becomes dry. When it reaches the ground, it turns and moves back towards the equator. This circular movement of air was later called the *Hadley cell*.

It is interesting that, like Howard and Beaufort, Hadley was not a professional meteorologist. He was a British lawyer who loved meteorology. It seems that a love for science and hard work can be enough to help a person achieve great results.

Reflection

1. Imagine you are Francis Beaufort, and journalists are interviewing you. Tell them why you decided to invent your wind scale and why the scale is important.
2. Can you remember a day in your life when it was very windy? Where and when did it happen? How strong do you think the wind was on the Beaufort scale? (You can find the full scale on the internet.) Describe what the weather looked and felt like on that day.

6

MORSE, FITZROY, AND THE FIRST
MODERN FORECASTS

Meteorological science was progressing fast in the early 1800s. The discoveries and inventions made this progress possible. However, fishermen and farmers still depended on themselves to predict the weather. They used signs like the appearance of clouds or the behaviour of animals. For example, they used *weather-frogs*. Weather-frogs were tree-frogs that were kept in jars with some water at the bottom and a small ladder. People believed that if the weather was improving, the frog would climb the ladder, and if it was going to rain, the frog would go down the ladder.

The fishermen and farmers had to predict the weather themselves because there was no technology to deliver weather information to them. Everything changed with one very important invention: the electric telegraph.

Several scientists worked on developing the telegraph in mid-1830s, but it was New York University professor Samuel Morse who was most successful. He created the cheapest and most convenient telegraph. He also created the Morse code, which is a set of different sounds. Each sound represents one letter of the alphabet.

The telegraph and Morse code made it possible to send information over long distances in seconds. Before the telegraph, weather reports could not travel faster than 160 kilometres per day, and most of the time they were even slower—usually only 60-120 kilometres per day. Imagine receiving a forecast and realising that it is a forecast for yesterday! How helpful is learning about an approaching storm after the storm has already destroyed your house and crops? By the late 1840s, the telegraph made it possible to deliver information over long distances almost immediately. That made forecasts much more useful.

The father of modern weather forecasts was Francis Beaufort's friend Robert FitzRoy. Like Beaufort, FitzRoy had a brilliant career as a Royal Navy officer. He became famous for being the captain of the *Beagle* during Charles Darwin's voyage and also for being a governor of New Zealand. In his later years, he also became interested in meteorology. He wanted to help sailors and fishermen who sometimes died because they didn't know about approaching storms.

In 1854, FitzRoy managed a new department of the Royal Society, which collected weather data at sea. FitzRoy ordered captains of ships to provide information about the weather. The captains were given the instruments necessary to collect weather data. FitzRoy also invented a new type of barometer. These barometers were fixed at every port, and captains of ships could look at them before going out to sea. This information saved many lives.

In October 1859, a terrible event happened. The *Royal Charter* ship sank off the north coast of Anglesey in Wales. It was returning from Melbourne, Australia when it was hit by a storm. The ship was carrying 375 passengers, 112 crew members, and many boxes full of gold from the gold mines of Australia. Only 40 people survived after the *Royal Charter* hit the rocks. All the women and children on the ship died. The terrible news of the ship led FitzRoy to develop maps to help predict weather. He called his maps *forecasts*. When we talk about a forecast today, we are using the word invented by FitzRoy.

FitzRoy set up 15 weather stations on land that used telegraphs to send daily weather reports. After that, the gale warning service was created.

The first daily weather forecast in history was published in *The Times* newspaper on 1 August 1861. Later that year, *The Times* published the first weather maps. *The Times* forecasts became popular very quickly. Everybody wanted to read them: sailors, fishermen, organisers of country fairs and flower shows, horse-racers, and more. At last, information about the weather reached everyone.

Reflection

1. Do you think weather forecasts are just as important today as they were in the 1700s? Why or why not?
2. Do a little online research. Find the name of another ship that sank because of bad weather. Where and when did it happen? What kind of weather caused the ship to sink? (Note that the *Titanic* did *not* sink because of weather!)

PART 4

COLLECTING PRESENT AND PAST WEATHER DATA

A tornado

Weather stations made it possible to collect large amounts of meteorological data. New instruments of data collection appeared in the 20th century. Thanks to this progress, we now have vast amounts of information about current and historical weather. In this section, you will read about instruments for gathering weather data, about some extreme weather that was observed using those instruments, and about some extraordinary weather records registered around the world. You will learn why climate is different from weather, what types of climates exist, how the Earth's climate has changed through history, and how it is changing today.

7

WEATHER STATIONS

The weather stations set up by FitzRoy (see Chapter 6) were not the first. The world's oldest weather station stands in the mountains 60 kilometres away from Munich, Germany. It has been collecting meteorological data since 1781 and is still working. The first observations at the station were taken by monks (religious men) from the nearby Rottenbuch monastery (a home for monks). When the monastery was closed in 1803, the monks, priests and village teachers continued making observations. At first, they weren't paid for their work, but later they were employed by the Bavarian Academy of Sciences. In 1952, the weather station became the Hohenpeissenberg Meteorological Observatory.

Today, the World Meteorological Organisation (WMO) controls more than 10,000 staffed and automatic weather stations on the surface of the Earth, 7,000 on ships, 1,000 in the air, and 1,000 on water. There are hundreds of weather radar tracking systems and 3,000 airplanes that carry special equipment. All of them measure the weather every day, collect data about the atmosphere and the ocean surface, and help make forecasts. Every country in the world can get information about the weather through the WMO Information System.

People might think that meteorologists use weather data only from the big WMO-controlled weather stations. However, that's not true. Sometimes the data comes from ordinary people like you.

Any weather fan, school, or organisation that wants to know the weather can buy a personal weather station. The most advanced digital model costs about 300 US dollars, and the cheapest model only costs about $20. The weather station can be put it in your yard, and it will give you information about the temperature, wind direction and speed, humidity, pressure, and amount of rain. It can have a lightning *sensor* (alarm) to warn about dangerous weather. You can add a camera to the weather station and share the local weather live on the internet or the radio. Some personal weather station owners even report the data to local meteorologists to help them improve their forecasts.

In developed countries, there are a lot of weather stations. They provide huge amounts of weather data necessary for accurate forecasts. In many developing countries, however, there are fewer stations and not enough data, so the forecasts are less accurate there.

One place with only a few weather stations in a large area is Antarctica. Antarctica is not a welcoming land. The weather makes living and working there challenging. Most of the continent is covered with ice all year round, and it is extremely cold. The temperatures rise above freezing in summer only in some areas along the coast. The temperatures inland are the lowest on Earth. The winds can be very strong. It is not easy to travel there, and it takes a lot of effort to keep the weather stations working all the time, which is why some stations are open only in the summer.

The first meteorological station in Antarctica was built by the Scottish scientist William S. Bruce in 1902. Bruce sailed to Antarctica with a group of explorers. They wanted to reach the South Pole. Although they never reached it, they explored new lands and did some research. Their main achievement was putting the first weather station on the continent.

The exact place for the station was chosen by accident. Approaching the Antarctic coast in January 1902, their ship *Scotia* ran into some great fields of ice. They had to look for a place to land. The safest and closest place was Laurie Island of the South Orkney Islands. The ship needed repairs, and the explorers needed to rest. When they landed, they started building a hut from available materials. The result was Omond House, the first meteorological station. Later, Bruce gave the house to the Argentinian government, and Argentina still runs the weather station there today.

In the second half of the 20th century, many new stations were built across Antarctica. The most important ones are the Amundsen-Scott station at the South Pole, which is run by the United States, and the Vostok station in East Antarctica, which is run by Russia. Both stations were built in 1957 as part of the International Geophysical Year. These stations are in the most unusual place on earth—an icy mountain desert with very low humidity and precipitation.

Another famous station is Halley station near the Weddell Sea, which is run by the United Kingdom, and has been open since 1956. The Mawson station, run by Australia, was built in 1954 and is famous for observing winds. Extremely strong winds, sometimes over 160 kilometres per hour, blow from the high ice sheet towards the sea.

There are over 30 staffed weather stations and over 160 automatic weather stations in Antarctica today. Because that number is always increasing, there is hope that we will know more about the weather and climate of this extraordinary continent in the future.

Reflection

1. Imagine that somebody gave you a weather station as a present. Where would you put it? What information would you be most interested in collecting? What would you do with the data you collected?
2. Do you want to visit Antarctica? Do you think you would like to live or work at a weather station in Antarctica? Why or why not?

8

RADIOSONDES, RADARS, AND SATELLITES

The 20th century saw great progress in weather observations. The progress was possible because of new technologies, which are still used today.

Meteorologists understood that they needed to know the weather both near the earth and high above the earth to make accurate forecasts. But there was no way to put instruments up high enough to get the information they needed.

In the 19th century, meteorologists tried to use kites to measure the temperature and pressure above Earth. But the kites were tied to the ground, and they couldn't go very far up. They were also hard to control in windy weather. From the end of the 19th century until the 1920s, scientists experimented with balloons. Finally, the French meteorologist Robert Bureau flew the first weather sensor on a balloon in 1929. The sensor sent down weather data. Bureau invented the name for it: *radiosonde*. A year later, Russian meteorologist Pavel Molchanov designed and flew his own radiosonde. It reached a height of 7.8 kilometres and measured the temperature there:–40.7 °Celsius! Molchanov's design became popular around the world because it was simple and used the commonly known Morse code to send data.

Today, hundreds of radiosondes go into the air every day. Some are dropped from planes, but most are lifted by balloons. When a balloon reaches a height where the air is very thin, it explodes, and the radiosonde begins to fall slowly with a parachute. As it falls, it spends 60 to 90 minutes in the air, collecting meteorological data.

Another kind of technology that appeared in the 20th century was the *radar tracking system*. A radar tracking system is an instrument that uses radio waves to find objects. When a wave from a radar hits an object, the wave is reflected back from the object. The reflection tells the people operating the radar how far away the object is, where it is, and how fast it is moving. Radars were developed secretly in the United Kingdom and the United States during World War II. When the military started using radars, they noticed that the waves were reflected not only from ships and other objects but also from rain and snow. People then realised that radars could be useful in meteorology.

Weather reports on television or the internet use radar maps that show the weather in real time. The uncoloured areas on these maps are places without precipitation. The colour green shows areas with light rain. Yellow shows moderate rain. Orange shows heavy rain, and red shows very heavy rain or rain and hail. White or blue indicates snow, and pink indicates freezing rain or sleet.

Information from radars has saved many lives because it has been used to predict storms and hurricanes. At first, meteorologists did not share information from radars on TV. The first time TV audiences saw a radar map was in 1961 in the United States. A very large hurricane named Carla was approaching the state of Texas. Dan Rather, a local TV reporter, travelled to a radar site and convinced the staff to let him show the computer's black-and-white image of the storm live on TV. He then asked a meteorologist to put a hand-drawn picture of the Texas coast on top of the image, using a sheet of plastic that you could see through. The audiences saw the size of the storm, where the eye of the storm was, and in what direction it was moving. After that, about 350,000 people were ordered to leave their homes. Thousands of lives were saved, thanks to Dan Rather and the first radar image on TV. A smaller hurricane in the same area in 1900 had killed more than 10,000 people, but in 1961, just 46 died.

Since 1970, radars have been connected to one another to create systems. In the 1970s, the first Doppler radars appeared. Most radars today are Doppler radars. These provide information not only about a storm's location but also about all movements within the storm at any moment. They are particularly good for predicting tornadoes.

Another important kind of technology that appeared in the 20th century is *weather satellites*, which observe weather systems around the world from above the earth. There are about 160 weather satellites in space today, and they make about 80 million observations per day. The satellites measure rain, snow, ice, dust storms, and ocean currents. They also measure wind by observing cloud patterns and how the clouds change with time.

During the day, satellites take normal photo images, which are easy for any person to see and understand. They can also take *infrared* images, which are only possible at night and that only an expert can understand. Infrared images provide information about the type and height of clouds, the ocean's temperature, and the movement of warm air, such as the air in El Niño. The infrared information is important for fishing and farming, and for planes and ships. Infrared also gives information about the strength of hurricanes. The greater the temperature difference between the warm eye of the hurricane and the cold cloud tops around it, the stronger the hurricane is.

There are two types of weather satellites. The first type are *geostationary satellites*. These stay in the same place, very high above the equator, and move together with the earth. The second type are *polar-orbiting satellites*. These go around the earth from north to south or from south to north. They move at different speeds. It takes a satellite from 10-30 days to fly around the whole planet. They are much closer to the earth than geostationary satellites. Both types of satellites are very important for weather forecasting. Geostationary satellites provide greater amounts of data, but polar-orbiting satellites provide higher-quality images with a lot of details.

Finally, one more useful piece of technology is *weather buoys*, which float in the water. They replaced weather ships in the 1970s, and they collect weather data from the world's oceans. There are now about 1,250 weather buoys in the world.

Reflection

1. What are some advantages to using modern technology to forecast weather? What are some advantages to using traditional methods?
2. Imagine you are Dan Rather, and journalists are interviewing you after the 1961 hurricane. Tell them how you got the idea to travel to the radar site. Describe how you felt when you realised how big the hurricane was. Tell the journalists how you feel about your actions now.

9

WEATHER RECORDS: THE HOTTEST, THE COLDEST, THE RAINIEST

People are always interested in all-time records. Sport records, unusual records from the *Guinness Book of World Records*, and weather records are all fun to learn about. Records sound exciting because they are so different from what we see and experience in everyday life. We know about weather records thanks to the meteorological data from weather stations, radars, and other instruments. Weather records show how different the weather can be in different parts of the world and at different times of year.

The highest temperature ever measured on Earth was 56.7°C (134.1°F) in Furnace Creek, California, USA, on 10 July 1913. The coldest temperature on Earth was registered in Antarctica. On 21 July 1983, the thermometer at the Vostok Station hit -89.2°C (-128.6°F).

Outside Antarctica, the coldest day was 22 December 1991, recorded at the Klinck Automated Weather Station in the north of Greenland: -69.6°C (-93.3°F). It is not surprising that Greenland can get so cold, but there are other countries with very cold temperatures that might surprise you. Here are some record cold temperatures from countries we normally think of as warm (many of these records were set in the mountains, where the air is colder):

- Morocco: -23.9°C
- South Africa: -20.1°C
- Afghanistan: -52.2°C
- India: -48.0°C
- Japan: -41.0°C
- South Korea: -32.6°C
- Italy: -49.6°C
- Spain: -32.0°C
- Australia: -23.0°C

Some hot and humid places near the equator, on the other hand, have never reached anywhere near the freezing point. For example, the coldest ever temperature in Singapore was 19.0°C (66.2°F) on 14 February 1989. You can visit Singapore at any time of year and safely leave your coat or sweater at home.

The United Kingdom is known as a wet and cloudy country without major temperature changes during the year. However, it has seen some extreme weather as well. For example, the winter of 1962-1963 was known as the Big Freeze because the temperatures stayed below freezing for weeks. The whole country was covered with snow from the end of December until the beginning of March. The rivers and the sea along the coast froze. On some days, the winds reached 8 on the Beaufort scale, and piles of snow reached the height of six metres (20 feet) in some places. The Big Freeze of 2009-2010 didn't last as long as the previous Big Freeze, but the temperatures hit record lows: about -20°C (-4°F) in England, Scotland, and Wales.

The summer of 2022, on the other hand, was extremely

hot in the UK. There were three heat waves in June, July, and August. They were not very long, but they were unusually powerful. Many records were set. On 19 July, the temperatures in some places went above 40°C (104°F) for the first time in recorded history. During this time, it was also extremely dry, and people had to save water. The drought caused many fires across the country.

Weather stations also give us historical data about the fastest changes in temperature. The most extraordinary temperature change happened in the Black Hills area of South Dakota in the United States. Because of the many hills, this area is famous for strong winds. Sometimes warm air comes from above and drives away the cold air near the Earth's surface very quickly. On 22 January 1943, in the town of Spearfish, the temperature rose from -20°C (-4°F) to 7°C (45°F) in just two minutes. That's the fastest warming in the history of observation. Only a couple of hours later, the wind changed, and the temperature dropped from 12°C (54 °F) back to -20°C (-4°F) in 27 minutes.

Rain reports from weather stations tell us that the rainiest day happened in Cilaos, Réunion on 7-8 January 1966, during Tropical Cyclone Denise. In 24 hours, 1,825 millimetres (72 inches) of rain fell—as much as London normally gets in more than three years! The most rain to fall in one minute was 38 millimetres, in Grande-Terre, Guadeloupe between 11:03 and 11:04 a.m. on 26 November, 1970. London normally gets that much rain in the whole month of March. A drier place like Lima, Peru gets that much rain in six years. Just think: one minute of rain in one place on Earth is the same as another place gets in years! What an amazing planet we live on!

The differences between wet and dry weather can also be striking. In 1939-1940 in Hawaii, it rained non-stop for 331 days. The longest period without any precipitation lasted more than 14 years, from October 1903 to January 1918 in Arica, Chile. The deepest snowfall happened on 14 February 1927 on Mount Ibuki, Japan, when 11.82 metres (about 40 feet) of snow fell in one day.

The wind records are extraordinary too. Winds of 90 to 100 kilometres per hour are rare on land. They measure 10 on the Beaufort scale and can bring a lot of damage. It is hard to believe that the wind can reach a much higher speed, but these higher speeds can happen during tornados. The fastest wind ever recorded was 484 km/h (1,587 miles per hour). It was observed by a radar in the tornado near Oklahoma City in the United States on 3 May 1999. The tornado completely destroyed over 8,000 homes, injured about 600 people, and killed 36.

Winds can get almost as strong during tropical cyclones. The fastest non-tornado wind was 408 km/h (1,338 miles per hour). It was recorded in Cyclone Olivia, which passed over Barrow Island, Western Australia on 10 April 1996.

Reflection

1. Do some research online and find different weather records for your city or geographical area (heat, cold, wind, precipitation, etc.). What are they?
2. Which records were the most surprising to you? Which were the least surprising?

10

TYPES OF CLIMATES AND CLIMATE CHANGES

The *climate* is what the weather is *usually* like in a certain place. For example, the United Kingdom is wet and cloudy, and Lima, Peru is dry—so we say that the UK has a wet climate and Lima has a dry climate. If *weather* changes, it means a temporary change that happens over a short time. If the *climate* changes, it means a change that happens over many years or centuries. Changes can happen to local climates. For example, an area with normal humidity can slowly turn into a desert. Changes can also happen to the earth's climate as a whole.

There are different reasons why our planet's climate has been changing for millions of years. They include changes to the Earth's path around the sun, the sun growing hotter or colder, volcanic activity, large objects falling to Earth from space, and changes in ocean currents.

Millions of years ago, Earth didn't have ice caps at the North and South Pole. Dinosaurs walked around and ate leaves from trees where nothing grows today. They even lived in Antarctica. Earth's climate has become much colder since then. About 2.5 million years ago, the Ice Ages began and lasted until 11,500 years ago. During that time, the climate kept changing between very cold periods and very warm periods. In the last cold period, almost all of Canada and the northern half of Europe were covered with ice and snow.

More recently was the Medieval Warm Period (MWP) between 900 and 1300. This warming had a great impact on world history. The northern seas became free from ice. Because of that, the Norsemen were able to sail to Iceland, Greenland, and the New World (later called America). Today, Greenland sounds like a strange name for a land of snow, ice, and grey rocks, but it really was green at that time. The Norsemen could grow crops and raise cows and sheep because of the warm climate. There were even some forests in the southern part of the island. The living conditions improved in Europe, too. There was enough food because crops grew well. People started moving to the north of Europe and using new lands for farming. England was so warm that grapes grew there and people made wine. The population of Europe grew fast. All of those changes were positive.

At the same time, people's lives in some other parts of the world were getting worse. The Native Americans were especially unlucky. Horrible droughts happened in parts of North America and around the great Mayan cities of Central America. Whole cities in the Andes disappeared due to lack of water. Lake Titicaca stood empty, and there was not enough water in the rivers flowing from the mountains to the sea. At the same time, in northern, central, and western Australia, which was usually dry, the climate became wet, and there were frequent floods and storms. The local people faced great challenges because they could not hunt animals for food.

The Medieval Warm Period was followed by the Little Ice Age. It lasted from around 1300 to around 1850. Part of the

Little Ice Age happened during the age of modern meteorology, so we know about it not only from historical documents but from weather station data as well. The greatest cooling happened in the North Atlantic region. The 14th century was a terrible time for Europe. As the climate got colder and rainier, the summers were rarely warm, and crops didn't grow well. There were long periods of hunger and disease, and the population of Europe decreased.

Have you seen the 15th-18th century paintings by Dutch and other artists that show snow-covered hills and villages, and people skating on frozen rivers and lakes? That was the normal winter weather in Western Europe at the time. The river Thames frequently froze in the 17th century, and people enjoyed *frost fairs* on the river's icy surface. People danced, skated, played football, and raced horses on the Thames. But it wasn't all fun. Life continued to be hard for the poor populations of Europe, who didn't have enough money to eat and keep warm. Periods of hunger continued for centuries.

Far away, in China, the unusually cold climate of the 17th century was one of the reasons for the fall of the Ming Dynasty. There were regular droughts in some parts of the country and floods in other parts. The extreme weather destroyed crops and led to hunger. The government could not provide the army with enough food. Together with some political problems, the climate led to the end of the Ming Dynasty.

Since the middle of the 19th century, the Earth's climate has been getting warmer. At first the warming was natural, but later, human activity added to the problem, and temperatures have been increasing even faster in the 21st century. For many years, people have been burning fossil fuels—oil, coal, and natural gas. As a result, it is harder for the heat from the sun to leave the atmosphere, and a *greenhouse effect* is created

—gases in the Earth's atmosphere trap the sun's heat and make the planet hotter. There are more heat records now than cold records because of global warming. Countries and international organisations are taking the situation seriously. They are creating new laws to limit the amount of greenhouse gases in the atmosphere and slow down global warming.

In addition to the global climate, people study local climates in different parts of the world. In Yakutsk, Russia, for example, you will experience a *continental* climate. The weather is generally dry. A normal summer day is hot, with temperatures reaching 30°C (86°F). But in winter, the temperatures stay below -30°C (-22°F) almost all the time. In other places, the differences can be less extreme. The climate of Minnesota in the United States is also continental, but the temperature often rises above freezing in winter.

If you travel to Dublin, Ireland, you will find a typical example of *maritime*, or sea, climate. The seasonal changes are small, and it often rains. If you tell a friend that it was 12 °C when you were in Dublin, they won't know if you travelled in January or July.

The *Mediterranean* climate is also a kind of sea climate, but it has hot, dry summers and mild, often rainy winters. You will experience this climate in Rome, Italy; Cape Town, South Africa; and Santiago, Chile.

A *desert* climate is extremely dry. It can get very hot during the day and very cold at night. It is hard for people in live in desert climates. That is why the populations of the Sahara Desert in Africa and the central parts of Australia are very small.

If you travel to the rainforest of Brazil, you will experience warm, humid weather all year round. There are no seasonal changes of temperature, and there is never any ice or snow. This type of climate is *tropical*. Some other tropical areas, such as India and Bangladesh, have two seasons: wet and dry. It rains almost all the time during the wet season. Still other tropical areas, for example the *savannahs* of Tanzania in Africa, are dry most of the year.

Reflection

1. What type of climate do you live in? Do you like it? What climate would you like to live in? Why? What climate would you never want to live in? Why not?
2. Can you think of anything that people can do to slow down global warming? Conduct a quick online search for some ideas.

PART 5

METEOROLOGY SERVING PEOPLE: TODAY AND TOMORROW

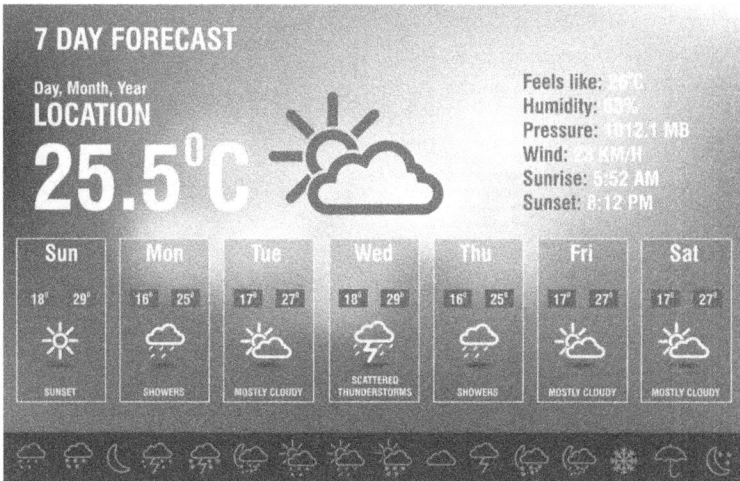

The weather forecast for one week

In this final section, you will learn how technological progress has driven meteorology forward over the last century. You will find out how people and machines work together today to create the best forecasts possible and what meteorologists say about the possibility of a 'perfect forecast' in the future. You will take a look at the life of a meteorologist and get some practical advice for what to pay attention to in a weather forecast.

11

LEWIS F. RICHARDSON, SUPERCOMPUTERS, AND MODERN FORECASTS

Many great ideas appear before there is technology to make them work. An example of this is *numerical* weather forecasting, which uses maths. More than 100 years ago, British scientist Lewis F. Richardson thought that human predictions of weather were not accurate enough. He suggested predicting weather by solving mathematical equations called *models*. This method is what meteorologists use today. But today, meteorologists have computers. A hundred years ago, Richardson had to imagine something else.

In his 1922 book *Weather Prediction by Numerical Process,* Richardson described a fantasy weather forecast factory. In his fantasy factory, there were 64,000 computers working in "a large hall like a theatre." He imagined that those computers calculated weather forecasts using data from weather balloons around the world. Are you surprised that Richardson imagined computers long before they appeared? Well, when he wrote "computers", he really meant people computing weather. He had no idea there would be machines called computers many years later.

Richardson imagined a huge world map on the walls of his

fantasy hall. Each 'computer' would sit next to a specific part of the map and calculate the weather for that part of the world. Special people would check the work of several nearby 'computers' to make sure their forecasts matched. Another person would watch and make sure everyone worked at the same speed. If someone was computing too fast, the person watching would shine a pink light on that region on the map. If someone was working too slowly, he would shine a blue light on that part of the map. Four clerks would collect information and take it to a special room. From there, it would be communicated to a radio station by telephone. Without realizing it, Richardson created a human model of a real computer of the future.

Why did Richardson believe 64,000 people were needed? Because without real computers, calculations took a long time. Richardson thought that 64,000 people would be necessary to complete the work quickly enough. With fewer people, the forecast would come too late to warn people about a storm.

To test his theory, Richardson tried to forecast the weather for a single date in the past—20 May 1910—using the weather data from 7 a.m. on that day. He knew what the weather really was later that day, so he wanted to see how accurate his numerical forecast would be. It took him years of difficult hand calculations using his mathematical models to finally produce a forecast. But the forecast was totally wrong. It looked like his efforts had failed. However, almost 90 years later, meteorologist Peter Lynch discovered the very small mathematical mistake Richardson had made in his calculations. Without that mistake, his forecast would have been accurate. It is clear that Richardson's work was ahead of its time. In fact, his approach is used in nearly all weather forecasts today.

When Richardson heard about the first weather forecast made by the first computer, ENIAC in 1950, he called the results an "enormous scientific advance." That 24-hour forecast took ENIAC nearly 24 hours to produce. Until the 1980s, numerical weather forecasts were not very reliable. After the 1980s, with the progress of computer technology, computer forecasts could be trusted more.

Weather forecasting in the 21st century is a good example of how people and machines can work together. It is hard for a human alone to make an accurate forecast. A forecast by a computer alone is not always accurate either. But when they help each other, the result is much better. Almost any forecast today begins with numerical calculations of pressure, temperature, wind, and humidity. The computer then uses models to produce a forecast. The human's job is to read that forecast and make corrections based on experience and knowledge. There are some websites that publish only computer forecasts. Such forecasts are usually less reliable than those checked by meteorologists.

As computer technology progresses, there is less and less work left for humans. Does that mean one day computers will replace human meteorologists completely? Probably not. For example, temperature models are almost perfect, and computers can usually predict temperature very accurately. However, precipitation is harder to predict. In the same area, it may rain or snow heavily in some places but not rain or snow at all in others. Local geography, such as mountains or lakes, can make a difference. A computer model may not handle that well, and human knowledge of the area is useful. Humans are also important in forecasting dangerous weather, such as hurricanes, tornadoes, or floods. A mistake can be very serious, so any machine forecast needs to be carefully checked by meteorologists.

Modern forecasting computers are called *supercomputers*. Their speed is unbelievable. They can do more than 10^{15} calculations per second. They are about six million times more powerful than the average desktop computer. A supercomputer receives large amounts of data from weather stations, Doppler radars, weather satellites, radiosondes, and other instruments. Meteorologists call such data *nowcast* because it describes the weather conditions right now. This information, in the form of numbers, goes into the supercomputer's numerical models. There are a lot of models, and they use enormous amounts of past and present weather data.

Different models might deliver different results because they give importance to different kinds of data. Some give more importance to wind speed, others to temperature or to humidity. Some models are better at predicting the weather for days ahead, while others are better at predicting for only hours ahead. If predictions from different models don't match, we learn about that from the weather forecast. The forecast may then give a broader temperature range. It may say, for example, that the afternoon temperature will be 'from 10 to 15' °C (or 'in the fifties' °F). It may also say that the chance of rain is 60% or 40% or 20%. Older forecasts used to sound more certain: "It will rain in Greenwich on Friday morning." Modern forecasts are more cautious. They talk about the *chance* of precipitation, which can be close to but almost never is 100%.

Each forecast model divides a geographical area into a number of squares. The forecast's accuracy depends on the size of the squares. The smaller the squares are, the more accurate the forecast is. But large squares can help predict general weather changes over longer periods of time. For example, they can show how a big storm will move across the continent over the next week. Together, these models create forecasts that save lives. Meteorologists can now predict almost exactly where a hurricane will hit land 24 hours before it happens. The average mistake is only about 70 kilometres (43 miles). Even five days before the event, they can predict the exact place with an accuracy of a little over 300 kilometres (about 190 miles).

Reflection

1. Richardson's work was ahead of its time. What does it mean if something is "ahead of its time"?
2. What are some other kinds of calculations that were harder before we had computers?

12

A DAY IN THE LIFE OF A
METEOROLOGIST

Most meteorologists work in offices, weather stations, or laboratories. They are surrounded by modern weather recording equipment; computers; and atmospheric, land, and water maps. Some meteorologists also work outdoors. University and government researchers may do field work to collect data or observe weather events. TV or radio meteorologists may report on weather events from the field (that means, from outdoors) as well as from a studio. These meteorologists used to be called weathermen and weathergirls. Today they are called weather presenters or weather forecasters. A weather presenter may not have an education in meteorology, but a weather forecaster must study meteorology in school.

Imagine you are a weather forecaster. What might your usual day be like?

You might wake up at 1 a.m. (Yes, you'll have to get up early if you want to be a meteorologist! The forecast must be ready before the rest of the country wakes up!) You'll get to work at around 3:30 a.m. to prepare the day's forecast. You'll look at satellite and radar data and at computer forecasts to make your own weather predictions. After an hour of studying data and making graphics, you'll be ready to appear on TV for the first time at 4:30 a.m. You'll give weather updates on news shows all morning. Your last update will be at 10 a.m., but your day won't end there. There are several options for what to do next. You might take a nap. Then you might visit local schools as part of a programme to teach children about the Earth, weather, and climate. You might continue studying new data and developing new forecasts in your office. You might teach a meteorology class at a local college or university. Whatever you do next, you'll have to go to bed early, at about 6 or 7 p.m., to be ready for the next day's early start.

While you're presenting a forecast on TV, your audience will see you standing in front of a moving weather map. In reality, you'll be standing in front of a blank green wall. Thanks to special effects, a digital map is put on top of the green wall for the viewers to see. You will see the map on a TV monitor, which the viewers don't see. That's how you will know where to stand and point. It is not as easy as it sounds. As a beginning presenter, it may take you a long time to learn to stand at the right place and use your hands in the right way.

While other TV presenters may read their text from a script on the screen, there are never any scripts for weather forecasters. You'll look at the changing map and use what you see to tell your stories. You'll have to think quickly and find the right words to describe the weather on the screen and to make predictions. A good TV meteorologist knows how to explain some very complicated weather events and scientific facts in simple language. Ordinary people have to understand what is going on in the atmosphere. According to Bob Henson, a meteorologist from the Weather Underground weather service, you need to "approach the weather as if you're telling a story: Who are the main actors? Where is the conflict? What happens next?" Henson says that when you present a forecast, you always do a bit of teaching. Weather forecasters are often the only scientists that viewers see regularly on TV.

More than 40 years ago, people's growing interest in weather gave birth to the Weather Channel. It first went on the air on 2 May 1982 in Atlanta, Georgia in the United States. It was the first television channel to give weather forecasts 24 hours a day, using a full staff of general meteorologists and specialists. The Weather Channel is still popular with American audiences, and some forecasters become local celebrities.

Not every meteorologist's job is to predict the weather, however. Some meteorologists are *atmospheric researchers*. They research how the atmosphere works and how it is connected with the water on our planet. They work together with other scientists studying complicated global issues like climate change.

Other meteorologists are called *climatologists*. They study climates of the past, work out patterns, and predict future changes. To find out about the weather since the 19th century, they use data from weather stations. Finding out about earlier periods is like the work of a detective. Climatologists use historical documents on food and wine production, the length of crop-growing seasons, and the length of time rivers and lakes stayed frozen to make conclusions about past climates. They also use physical objects, such as tree rings. They check ice from Greenland, Antarctica, and high mountains to find remains of ancient snow storms. They study the oceans' bottom to find evidence about weather conditions from different periods of the past. Today, most climatologists study global warming.

Still other meteorologists are called *forensic meteorologists*. They study weather conditions at specific moments in the past; for example, when a crime or an accident happened. They try to find out how much the weather influenced that crime or accident. They often bring their information to court to help judges and lawyers.

Finally, there are meteorologists who develop and improve weather technology and computer models. They may also design digital weather displays.

Reflection

1. What do you think are the most interesting and exciting parts of being a meteorologist? What do you think is the most difficult part of the job? The most boring part?
2. Would you like to be a climatologist or forensic meteorologist? Why or why not?

13

THINGS TO KNOW ABOUT WEATHER FORECASTS

Sometimes people complain about wrong predictions made by meteorologists because they don't know a few simple facts about forecasts. If people knew those facts, they might complain less.

The first fact is that the weather inside a city may be different from the weather in the surrounding countryside. The air is usually warmer in the city centre. The real temperature in the city centre may be several degrees higher than in the forecast. In winter, cities are warmer because human activity, cars, and buildings produce heat. In summer, cities can be warmer because the streets and parking lots get hot in the sun and heat the air above.

Luke Howard, the Godfather of Clouds (see chapter 5), was the first to explain city weather in his 1820 book *The Climate of London.* He described a 'heat island' effect, when the nighttime temperatures in London were 2.1°C higher than in the countryside. However, the daytime city temperatures were actually lower than in the countryside. This happened because of the *smog*, which he called 'city fog'. The smog hung over the city and blocked the heat from the sun. Today, smog is rare in London, so the temperature in the centre of London is usually higher than the temperature outside the city both at night and during the day.

The wind in cities is hard to predict too. Between two tall buildings, the wind may be very strong, and the air may feel cold. But if the buildings block the wind, a windy day might seem like a calm one.

The second fact is that weather stations always measure temperature in the shade, away from direct sunlight, and 1.25-2 metres above the ground to find the real air temperature. If a thermometer is in the sun, or has recently been in the sun, it may show a much higher temperature than meteorologists report. It doesn't mean they are wrong. If the temperature was measured in the sun and on the ground, all heat records would be 30-50 degrees higher. For example, the ground surface temperature at Furnace Creek, California on 15 July 1972 was 93.9°C (201.0°F). Be careful not to walk around without shoes on such a hot day!

A mistake people make when they read forecasts is that they look only at the temperature but not at what it *feels like*. Think about stepping outside in the morning and thinking, "I knew it would be cold—but not *this* cold!" Many people miss the 'feels like' part of the morning's weather report. Today, weather reports usually tell viewers both what the temperature is and what it feels like. Forecasts also include 'feels like' predictions next to the predictions of the real temperature. A viewer might see, for example: "temperature 3°C, feels like -10°C."

How do meteorologists calculate that 'feels like' temperature? The way you feel depends not only on real temperature but also on wind and humidity. A strong wind can make the air feel much colder than it really is. A humid day in summer can feel much hotter than it really is. Why is that? The water in your body evaporates through our skin all the time. On windy days, it evaporates faster and moves heat away from the body. When that happens, you feel like it is colder than it really is, especially in winter. On hot and humid days, the water evaporates more slowly. When that happens, there is no cooling, and you feel like it is hotter than it really is. Meteorologists use a special formula to calculate the 'feels like' temperature based on their understanding of how the human body works.

Sometimes people complain about forecasts because they expect them to remain accurate for many days and weeks. They may check a ten-day forecast and trust its prediction for tomorrow as much as its prediction for ten days ahead. However, forecasts lose accuracy over time. With each new hour and day, the forecast becomes less and less reliable. Predictions for the next 24 hours are always more accurate than for the 24 hours after that. Predictions for the day after tomorrow are always more accurate than for five days ahead. Predictions for ten days ahead come true as often as they don't. So don't be upset if you want to plan a picnic in ten days, but the forecast predicts rain and cold. Check it again next week, and it is possible that the updated forecast will make you happy.

The good news is that the accuracy of forecasts has been improving by one day in 10 years. To give an example, 20 years ago, a 3-day forecast was 90 percent accurate; 10 years ago, a 4-day forecast was 90 percent accurate; and in the early 2020s, a 5-day forecast was 90 percent accurate. We can expect that in the early 2030s, a 6-day forecast will be 90% accurate. Of course, this accuracy is achieved where the most modern technology is available. The forecasts in some countries and regions might be less accurate than in others.

But how much better can forecasts get? Can the improvement continue forever?

Reflection

1. Find and save the forecast for your area for two days ahead, five days ahead, and ten days ahead. Include predictions about the temperature, 'feels like' temperature, clouds, precipitation, humidity, pressure, wind direction, and wind speed. Then check the actual weather in two, five, and ten. How accurate was the forecast? What was incorrect?
2. Compare forecasts from different sources and see which source gives the best forecasts. Why do you think they might differ?

14

THE BUTTERFLY EFFECT: IS A PERFECT FORECAST POSSIBLE?

Scientists keep discovering new laws of the atmosphere. The amount of weather data available to them is always increasing. It includes both past and present data. Supercomputers are getting more and more powerful. Does it mean that a perfect forecast, which is never wrong, is possible?

The atmosphere is very complex. Therefore, creating a perfect model of it is difficult and probably impossible. People can *predict* what will happen in the atmosphere, but they can never be completely *sure* of it.

Have you heard about the *butterfly effect*? It is the idea that a single *flap* (movement) of a butterfly's wings in Brazil can start a tornado in Texas. In other words, a very small change to the initial weather conditions can lead to a very different result. Mathematician and meteorologist Edward Lorenz introduced the idea of the butterfly effect in 1972. He gave it that name after reading the 1952 science fiction story "A Sound of Thunder" by American author Ray Bradbury. In Bradbury's story, the hero travelled to the past and accidentally killed a butterfly there. When he returned to the present, he realised that this small accident had completely changed the course of history.

While using a mathematical model to predict weather, Lorenz noticed something strange. When he made very small changes to the initial conditions, he sometimes got very different predictions even with the same model. He compared those small changes to a flap of a butterfly's wings. He noticed that even the tiniest flap could result in a big storm. Without it, the storm would never have happened.

Lorenz understood that nobody could predict every little change in the atmosphere. He concluded that there was a mathematical limit to any predictions of weather. He thought the limit was two weeks.

Although technology has improved greatly since the 1970s, Lorenz's conclusion is still generally true. Falko Judt, a research meteorologist at the National Center for Atmospheric Research in Boulder, Colorado, gave an interview in *Discover* magazine in 2019. He said, "We will never be able to actually achieve perfect forecasts." According to Judt, it will never be possible to predict the exact time and place of a thunderstorm more than two hours before it. For bigger systems like hurricanes and winter storms, he thinks the prediction limit is two to three weeks because people cannot measure everything at every point in the atmosphere all the time. We cannot watch every butterfly on Earth all the time.

Bob Henson of Weather Underground says, "The most salient [easy to notice and remember] weather forecast is the one that was wrong—when you expected something and you were surprised, those are the ones you remember. You don't remember all the times that it was just as we expected, because that's not news." Obviously, the main goal for meteorologists is to produce as many forgettable forecasts as possible.

Reflection

1. "Never say never" is a popular saying. Sometimes people are sure that something will *never* be possible. Then, suddenly, a big discovery or invention makes it possible. Do you think that Lorenz was right that there will *never* be a perfect forecast? Why or why not?
2. Ray Bradbury wrote about time travel. Do you know of any other stories about time travel? Explain what happens in the story.

15

ARTIFICIAL INTELLIGENCE AND THE FUTURE OF WEATHER FORECASTING

Now we know that there are limits to forecasting, could there be any surprises ahead? Could the latest technology make a difference? Artificial intelligence (AI) has recently become a popular subject of conversation. Suddenly, things people once thought were impossible are becoming possible. What could that mean for the future of weather forecasting?

AI systems do not need any models. They are different from numerical forecast systems. In numerical forecasts, a machine solves a great number of equations which scientists put into it. These equations are based on scientific knowledge of the laws of physics. AI does something different. A huge amount of past weather data is put into an AI system. This data can be for any geographical location or area. The AI system 'learns' from the data and develops its own understanding of how weather conditions change. Then, using present weather data from different observation instruments, AI makes predictions based on what it has learned about the past. Importantly, AI never stops learning and improving itself. It can get feedback on the accuracy of its forecasts and learn from that feedback. It will learn from its mistakes and might be right the next time in similar conditions.

It is possible that someday, such machine learning models will completely replace traditional numerical weather prediction models. If they do bring the expected results, they will make the work of meteorologists much easier. It won't be necessary to think of complicated equations for machines to solve. The machine learning models will use only a small part of the computing power that numerical models require. It won't be necessary for researchers to know programming languages. Even a researcher without this knowledge will be able to create good machine learning programs.

Several AI forecasting systems are already available. For example, DeepMind is especially good at predicting precipitation and other weather changes in the next two hours. NowCast is good at making predictions for longer periods, and it constantly changes them when it receives new data. A system called GRAF gets weather information not only from stations, radars, and satellites but also from sensors on planes and even from smartphones around the world when the users give permission.

As research shows, machine-learning weather systems can already predict general weather patterns as accurately as numerical models. However, they are not too reliable for short-term forecasts yet. There is no doubt that technology will keep improving. Will it ever prove Lorenz wrong about the butterfly effect and the impossibility of a perfect prediction? Time will tell.

At present, there are a few issues with AI predictions. The forecasts are made from past data, but the AI systems don't know physical laws. As a result, they may sometimes produce forecasts that are impossible in the real world—for example, predicting -40°C in Paris, France. It is also unclear how they will perform during weather events that have never happened before.

There is also the risk that rich areas of the world will benefit from the new technology more than poor areas because the rich areas will have collected much more past weather data.

Hopefully, AI and machine learning will help human meteorologists do their job more efficiently, so they can spend less time preparing routine forecasts and more time communicating forecasts to people. Meteorologists will not only explain how strong the wind will be or what the temperature will be, but also what the weather *might* do and how people can protect themselves from dangerous weather events. Humans and machines should work together towards common goals–to understand, inform, and protect.

However, there is one thing that only humans can do. It is what Aristotle did when he tried to understand the origins of clouds and rain. It is what Fahrenheit did when he created a new way of measuring temperature. It is what Beaufort did when he invented his weather codes and wind scale. It is what Richardson did when he described his forecasting factory— the model of a future computer.

Only humans can *imagine*.

Reflection

1. Can you think of any dangers of using AI to predict the weather?
2. No matter how clever a machine is, there are some things only a human can do. List a few things that only humans can do now. Do you think that machines might be able to do them in the future?

ABOUT THE AUTHOR

Alex Semakin is an English teacher and teacher-trainer with almost 30 years of experience and an MA TESOL from the University of Manchester. While continuing to train teachers of English, Alex has recently branched out into materials writing. He has written several coursebooks for Oxford University Press (China). He has also written and edited teacher-training courses for Language Fuel in New Zealand and INTESOL in the UK. Meteorology has been Alex's hobby since childhood. He began his first weather diary at age seven. His eyes are still fixed on the sky.